Investigating Science

GENETICS
The Study of Heredity

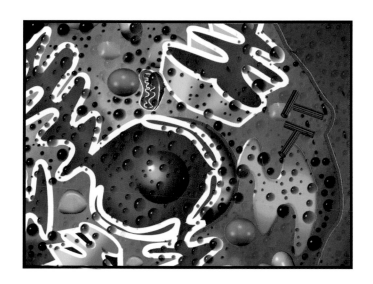

by Ian Graham

Gareth Stevens Publishing
A WORLD ALMANAC EDUCATION GROUP COMPANY

CONTENTS

Please visit our web site at: www.garethstevens.com
For a free color catalog describing Gareth Stevens Publishing's
list of high-quality books and multimedia programs, call
1-800-542-2595 or fax your request to (414) 332-3567.

Library of Congress Cataloging-in-Publication Data

Graham, Ian.
   Genetics: the study of heredity / by Ian Graham.
       p. cm. — (Investigating science)
   Summary: Examines the field of genetics, discussing DNA research, heredity,
cloning, genetic engineering, and the possibility of gene therapy.
   Includes bibliographical references and index.
   ISBN 0-8368-3231-0 (lib. bdg.)
   1. Genetics—Juvenile literature. [1. Genetics.] I. Title. II. Series.
QH437.5.G73   2002
576.5—dc21   2002022535

This edition first published in 2002 by
Gareth Stevens Publishing
A World Almanac Education Group Company
330 West Olive Street, Suite 100
Milwaukee, WI 53212 USA

This U.S. edition © 2002 by Gareth Stevens, Inc. First published by
ticktock Publishing Ltd., Century Place, Lamberts Road, Tunbridge Wells,
Kent TN2 3EH, U.K. Original edition © 2001 by ticktock Publishing Ltd.
Additional end matter © 2002 by Gareth Stevens, Inc.

Illustrations: John Alston, Simon Mendez
Gareth Stevens editor: Jim Mezzanotte
Cover design: Katherine A. Goedheer
Consultants: Angela Mengelt, MS, LuAnn Weik, MS, CGC, Genetic
Counselors, Children's Hospital of Wisconsin

Printed in Hong Kong

1 2 3 4 5 6 7 8 9 06 05 04 03 02

GENETICS
The Study of Heredity

CHARLES DARWIN

British scientist Charles Darwin studied many plants and animals during a voyage around the world in the 1830s. Darwin noticed how well-adapted plants and animals were to the places where they lived, and he concluded that living things changed over many generations to become better suited to their particular environments. He called this process "natural selection." In 1859, Darwin published *On the Origin of Species*. In this book, he introduced a theory called evolution to describe the way organisms changed over time. He could not, however, explain how these changes happened.

STUDYING ICELANDERS

In addition to examining tiny pieces of plant and animal tissue, geneticists also study whole populations. For example, geneticists often study the people of Iceland, who are all descended from one small group of settlers. This common set of ancestors allows geneticists to study how characteristics have been passed down through many generations of families.

WHAT IS GENETICS?

Genetics is a branch of science that deals with how the characteristics of living things get passed down through the generations. In all organisms, genes are responsible for this process, which is called heredity. Genes make up the chemical blueprint of life. They determine our appearance, and they can also determine how we behave and even what kinds of illnesses we will get. By studying genes, scientists are learning how our bodies work and even how to cure inherited diseases. Scientists who study genes are called geneticists.

SELECTIVE BREEDING

Farmers have used genetics for thousands of years to change certain characteristics of cows (*above*) and other livestock. By allowing only the finest animals to breed, farmers have ensured that the best physical characteristics have been passed down through the generations, so that dairy cows now produce more milk and cattle provide more meat. Controlling the way animals reproduce, rather than letting nature take its course, is called selective breeding.

TIME CAPSULES

Tens of millions of years ago, flies buzzed around the prehistoric world. Sometimes, a fly got trapped inside gooey sap oozing from a tree trunk. The sap then hardened and became amber (*right*). If the fly's last meal was blood sucked from a dinosaur, then the blood and the genetic information it contains might still be preserved inside the fly. In the film *Jurassic Park*, scientists brought dinosaurs back to life using samples found in these amber time capsules.

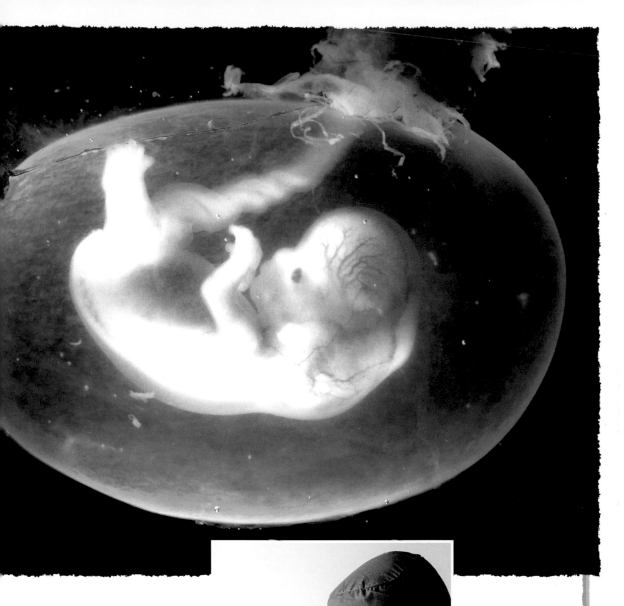

A NEW CRIME FIGHTER

Until fairly recently, a test of blood type was one of the few scientific tools available to police for solving crimes. Now, however, genetics has revolutionized crime fighting. DNA, which is the chemical that contains genetic information, can be retrieved from skin, blood, or hair found at a crime scene. An analysis (*right*) of this DNA can often lead to the arrest and conviction of a suspect. If DNA found at a crime scene is identical to a sample taken from a suspect, the chances of a wrong match are literally millions to one.

HEREDITY

Even when we are at a very early stage of development (*above*), many of our physical characteristics have already been decided. Genetic instructions that we inherit through the genes of our parents determine what we will look like and how we will grow. In a baby, even characteristics such as height have already been established.

MICRO-PIPETTES

Geneticists use tools called micro-pipettes to move very tiny amounts of liquid. When a button on top of the tool is pressed, a measured amount of liquid gets sucked up into the tip. Pressing the button again pushes the liquid out. The tip is replaced with each use to ensure cleanliness. Some micro-pipettes can be adjusted to draw up different amounts of liquids.

TRICKS & TECHNIQUES

Because geneticists often work with tiny pieces of plant and animal material, they need special tools to be able to do their jobs properly. Optical microscopes show tiny samples in greater detail, while the more powerful electron microscopes allow geneticists to study the structures inside cells. Geneticists use special tools, called micro-tools, to handle genetic material. Finally, they depend on a variety of equipment to help them analyze and understand the genetic information they uncover.

OPTICAL MICROSCOPE

An optical microscope (*above*) uses light to magnify samples up to 1,000 times their original size. It has an objective lens that forms a magnified image of the sample, and a second lens, called an eyepiece, for viewing the image. An optical microscope's magnifying power can be calculated by multiplying the magnifying powers of the two lenses. If a microscope has an objective lens that magnifies 40 times and an eyepiece that magnifies 10 times, it will make samples appear 400 times bigger. An optical microscope usually has at least three different objective lenses, so that different magnifying powers can be selected.

ELECTRON MICROSCOPE

An electron microscope (*right*) can magnify objects up to one million times their original size. It can show the inside of a cell in sharp detail. Instead of light, it uses beams of particles called electrons to form an image. The electrons either bounce off the sample or pass through it, and then they hit a screen that turns them into light. The image can be displayed on a computer screen.

DNA SEQUENCER

Scientists who study genetic information use a computer-controlled machine called a sequencer. It automatically works out the order, or sequence, of the chemical units in DNA, the substance that contains our genetic code. The sequence is vital for understanding how this genetic code controls people's bodies.

MICRO-TOOLS

Geneticists sometimes want to inject material into a cell, so they need micro-tools that are small enough to hold a cell without causing any damage. Extremely fine needles are also needed, with tips narrow enough to fit inside a single cell. Geneticists usually hold a cell in place by sucking it onto the end of a fine glass tube (*above*).

COMPARING SAMPLES

Scientists use a process called electrophoresis (*right*) to compare genetic material from different people. With this process, scientists produce special photographs of genetic samples, called autoradiographs, that show a ladderlike pattern of bands for each sample. By studying similarities and differences between the patterns, scientists can tell whether the samples came from people who are related to each other.

ELECTROPHORESIS

When using electrophoresis, scientists squirt a liquid containing pieces of genetic material into a special gel. An electrical current flowing through the gel causes the genetic material to move. Short pieces of genetic material travel farther than long pieces because they can move through the gel more easily. To create an autoradiograph, scientists add a special substance to the genetic material to make it radioactive and lay a sheet of photographic film on the gel. The genetic material darkens the film, forming a ladderlike pattern that shows how far the fragments of genetic material have traveled. Scientists can then compare the pattern with other examples.

WE ARE ALL UNIQUE

About six billion people live in the world today, but it is very rare to find two people who look exactly the same. Many factors determine appearance, and tiny differences in just one or two of these factors can distinguish people from each other. Even people who look alike may be very different in ways that we cannot see. For example, they may have different blood types or be affected by different medical conditions. Even if two people appear identical, scientists can still tell them apart.

A FACE IN THE CROWD

Consider the faces in a crowd of people. The faces are roughly the same shape and size, and they all have two eyes, two ears, a nose, and a mouth, yet each face is distinctive. We inherit unique characteristics from our parents that make us look different from each other. Our brains are good at spotting these differences.

THE GENETIC CODE

We are different from each other because the genetic information, or code, that tells our bodies how to grow and develop is incredibly complicated. With the exception of identical twins, the chance of two people having exactly the same genetic code is about the same as the chance of two people each tossing a coin thousands of times and coming up with exactly the same results.

RETINAL PATTERNS

The retina is a light-sensitive layer at the back of each eye. As an eye develops, blood vessels grow across the retina (*left*). The pattern the blood vessels make is unique to each person and can be used for identification purposes. Retinal identification works by firing an invisible infrared beam into the eye and picking up the reflection of the vessels from the retina. This pattern is then compared to patterns already stored in an "image" bank.

IDENTICAL TWINS

A woman normally produces one egg at a time. If it is fertilized by a man's sperm, it will develop into a single baby. If the fertilized egg splits into two, however, each part can develop into a baby, and the babies will be identical twins. Sometimes a woman produces two or more eggs at the same time. Each egg has the potential to develop into a baby. The babies that develop from these separate eggs, however, will not be identical to each other — they will be what are called fraternal twins.

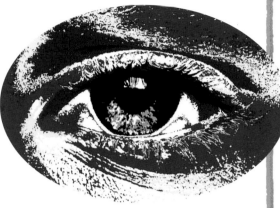

EYE ON THE FUTURE

Our eyes can be used to identify us because, like fingerprints, they contain unique patterns. In addition to retinal identification, a technology has emerged that uses the iris for identification. The iris, which is the colored part of the eye that surrounds the black pupil, is made from a ring of muscle. The patterns and colors in the muscle are unique to each person. Some banks are already testing new cash machines that identify people before issuing money to them by scanning their irises and comparing the results to images stored in a computer.

FAMILIES

Look at any family and you will probably notice that the children resemble at least one of their parents. Sometimes, however, children look more like their grandparents than their parents, because some characteristics do not show up in every generation. Red hair, for example, can be an inherited trait, but it may skip a generation before the next person with red hair is born into a particular family.

FINGERPRINTS

We all have patterns of raised lines on the ends of our fingers. These lines form before we are born and stay with us throughout our lives. People have been marking important documents with their fingerprints for thousands of years. Because every person's fingerprints are unique, the police can use them to identify people.

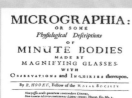

MICROGRAPHIA:
OR SOME
Phyſiological Deſcriptions
OF
MINUTE BODIES
MADE BY
MAGNIFYING GLASSES.
WITH
OBSERVATIONS and INQUIRIES thereupon.

By R. HOOKE, Fellow of the ROYAL SOCIETY.

LONDON, Printed by Jo. Martyn, and Ja. Alleſtry, Printers to the
ROYAL SOCIETY, and are to be ſold at their Shop at the Bell in
S. Paul's Church-yard. M DC LXV.

THE CELL PIONEER

In the 17th century, British scientist Robert Hooke became one of the first people to examine living things through a microscope. He built his own microscope, and in 1665 he published a book, *Micrographia*, with drawings and descriptions of his findings. Hooke was the first person to use the word "cell" to describe the walled bags of material he saw inside living things.

CELLS & THEIR STRUCTURE

All organisms are made up of tiny units called cells. A cell consists of two parts, the nucleus and the cytoplasm, with the nucleus controlling what happens in the cell. Each cell in an organism has a specific purpose. The flowers, roots, and leaves of plants are all made from specialized cells, and so are the muscles, bones, and skin of animals. The simplest living things on Earth consist of just one cell, while the most complex organisms, such as humans, are made of trillions of cells.

STORY OF A CELL

In an animal cell (*right*), DNA inside the nucleus determines which proteins the cell makes. Proteins carry out important functions in the cell. They are made from substances that enter the cell through its outer membrane. The proteins are formed in small ball-shaped structures called ribosomes, and they are stored in folded layers of membranes called the endoplasmic reticulum. Sausage-shaped structures called mitochondria break down substances to release the energy the cell needs, and a folded structure called the Golgi complex stores substances and moves them around to where they are needed.

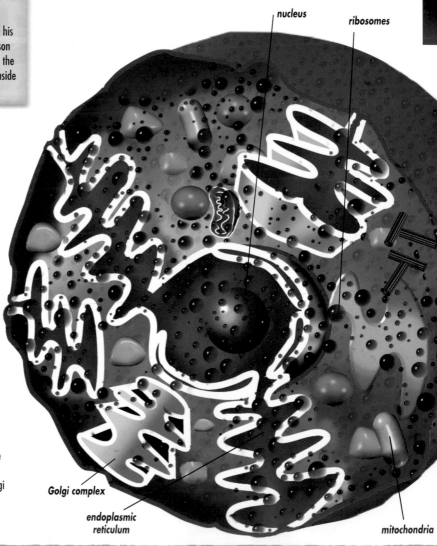

nucleus

ribosomes

Golgi complex

endoplasmic reticulum

mitochondria

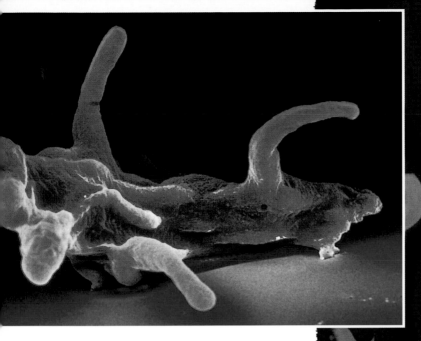

A SIMPLE LIFE

The simplest organisms, such as the *Amoeba proteus* (*above*), are made from only one cell, while larger, more complex organisms are made from many cells working together. The first cells that appeared on Earth more than 3 billion years ago had no nucleus, so their DNA was distributed throughout the cytoplasm. Cells with a nucleus evolved about a billion years later, and the first organisms with more than one cell appeared about 550 million years ago. Most animal cells (including those of humans) have one nucleus, but red blood cells have no nucleus.

MITOCHONDRIA

Mitochondria are tiny chemical factories inside cells that break down sugars to release energy. They contain genetic material inherited only from females. As this material was copied through the generations from one female to another, it sometimes changed because of copying mistakes called mutations. By studying these mutations in the mitochondria of various organisms, scientists can learn how and when organisms evolved, or changed over time.

CELL TYPES

Humans have about 200 different types of cells, including nerve, muscle, bone, skin, and blood cells. Blood cells (*right*) take in food particles and break them down into smaller, simpler substances. The blood cells then use these substances to make proteins that build and repair cells and take part in the many chemical reactions that go on throughout the body.

THE DNA LADDER

When DNA copies itself, it resembles a ladder that splits in half, right down the middle of every rung. Each rung of the DNA ladder consists of two bases that join together in the middle of the rung. DNA has four kinds of bases, known as T, A, C, and G for the first letters of their chemical names. T always connects to A, and C always connects to G. To create more genetic material for new cells, DNA copies itself by first splitting into two halves. Each half then "grows" back the missing half, so that two sets of DNA are produced from the original, single set.

THE DOUBLE HELIX

A DNA molecule has a shape, called a double helix (*above*), that resembles a twisted ladder. It is made of billions of chemical units called nucleotides linked together like beads strung out along a string. Each nucleotide consists of three parts — a chemical called a phosphate, a sugar called deoxyribose, and a chemical called a base. The phosphates and sugars form the two long intertwined strands of the double helix and the bases link these two strands together.

MAKING THE DNA DISCOVERY

In 1953, British scientist Francis Crick (*left*) and U.S. scientist James Watson (*right*) discovered the structure of the DNA molecule while they were working together at Cambridge University in England. They won the Nobel Prize in 1962 for their findings. The discovery of DNA was a groundbreaking achievement that enabled scientists to better understand the makeup of living things.

A CLOSER LOOK AT DNA

Plants and animals grow and function because of the amazingly complicated instructions in the genetic material inside their cells. This genetic material is called deoxyribonucleic acid, or DNA. Most DNA is found inside a cell's nucleus. Some DNA and a different sort of nucleic acid, called ribonucleic acid, or RNA, are found outside the nucleus in the rest of the cell. Molecules are groups of atoms, tiny particles that make up all substances. Because a DNA molecule can copy itself, the new cells that a growing organism produces all contain copies of the same set of genetic instructions, or genetic code.

SEX CELLS

There are two types of sex cells — the female ovum, or egg, and the male sperm. Both kinds of cells contain half the DNA found in other cells. When a sperm fertilizes an egg, their DNA combine to form a new cell with the correct amount of DNA.

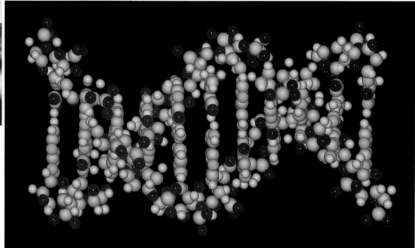

COILS OF COILS

The DNA molecule (*left*) is twisted into a coil, like a piece of string wrapped around a pencil. This coil is itself twisted into a coil. If the DNA found in just one human cell was uncoiled and stretched out in a straight line, it would measure more than 10 feet (3 meters) long. If all the DNA in your body was unraveled, it could stretch to the Sun and back again 600 times!

A DNA PIONEER

In the early 1950s, British scientist Rosalind Franklin provided a breakthrough that would enable other scientists to discover the structure of DNA (*opposite page*). She produced patterns on photographic film by firing X-rays at DNA. X-rays normally travel in straight lines, but DNA bends them. The bent X-rays made a pattern of spots on the film (*right*). These patterns gave other scientists vital clues about the structure of DNA.

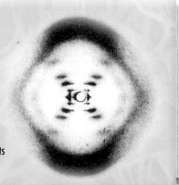

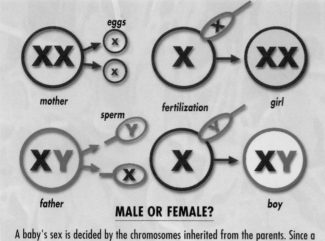

MALE OR FEMALE?

A baby's sex is decided by the chromosomes inherited from the parents. Since a woman's cells contain two X chromosomes, she can only pass on an X chromosome to the baby. A man's cells have an X and a Y chromosome, however, so he can pass on either an X or a Y chromosome to the baby. If a man passes on an X chromosome, the baby will be a girl. If he passes on a Y chromosome, the baby will be a boy.

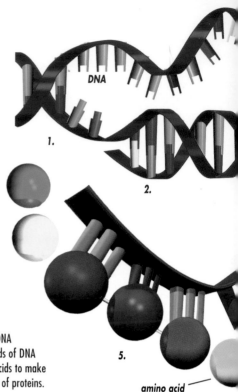

MAKING PROTEINS

When a cell creates protein, the two strands of DNA separate (1). A single strand of DNA produces a strand of RNA (2). The RNA "unzips" from the DNA (3) and the two strands of DNA come together again (4). The RNA joins together a series of chemicals called amino acids to make proteins (5). Different genes make different RNA, which in turn make different kinds of proteins.

amino acid

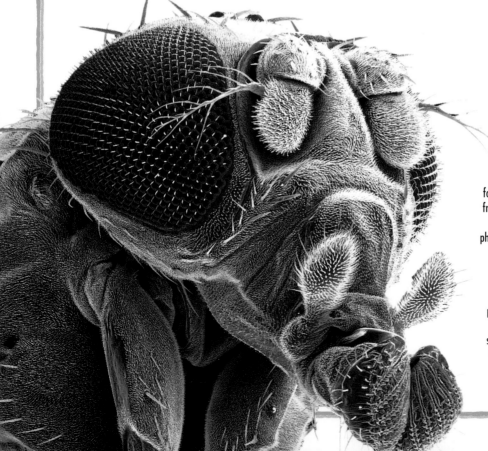

FAMOUS FLY

Because the fruit fly (*left*) has a simple genetic code, it was used in early genetic research at the beginning of the 20th century. Each cell of a fruit fly has only four pairs of chromosomes. When fruit fly genes mutate, or change, some mutations cause obvious physical changes such as strangely shaped eyes or wings. Since scientists only had to study four pairs of chromosomes, they could easily match each physical change to a mutation in the fly's chromosomes. The scientists were able to determine which part of each chromosome was responsible for each of the fly's physical characteristics.

CHROMOSOMES & GENES

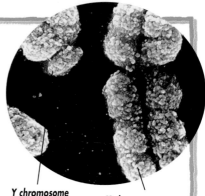

Y chromosome

X chromosome

A cell's DNA is packaged in short lengths known as chromosomes. Each chromosome consists of a series of smaller units of DNA called genes. Some genes instruct cells how to make the proteins that are essential for growth and the proper functioning of a cell, but scientists do not yet know what most genes do. Not every cell has all of its genes "switched on" all the time. In muscle cells, for example, only the genes responsible for making muscle proteins are switched on. Each human cell has 23 pairs of chromosomes, for a total of 46.

4.

3.

RNA

PAIRING UP

Chromosomes come in pairs, with one chromosome inherited from each parent. A female human has two "X" chromosomes that look alike, but a male human has an "X" and a "Y" chromosome (*above*) that look very different from each other.

BOYS WILL BE BOYS?

In humans, all male characteristics are due to only one gene, called the SRY gene, found in the Y chromosome. A baby will be a girl unless the SRY gene is present. This gene switches on other genes that produce certain chemicals, such as testosterone. These chemicals will change an embryo (a baby that is just beginning to grow) into a male.

CELL DIVISION

Cells do not last forever. Every hour, about 200 billion cells in your body die and have to be replaced with new ones. Cells increase their numbers by duplicating and dividing. First, each chromosome in a cell duplicates to form chromatids, which are two strands of DNA. These chromatids are pulled apart by fibers called spindles and are shared by two new nuclei before two new cells are formed. The chromatids then become the chromosomes for the two new cells.

PASSING ON OUR GENES

Organisms pass on characteristics to the next generation through copies of their genes. Some genes disappear through the process of natural selection, but many others have been passed down from generation to generation for thousands of years. For example, scientists have determined that the "ginger gene," which produces red hair, fair skin, and freckles, probably dates back to before the emergence of modern humans in Europe about 40,000 years ago.

ROYAL GENES

Genes often can be traced through royal families because royal ancestry has usually been recorded accurately for hundreds of years. Queen Victoria of England (*above, center*) inherited a gene that causes a condition called hemophilia. If people who have hemophilia get cut, bleeding does not stop right away because the blood takes a long time to clot. Hemophilia only affects men, but the gene that causes it is passed down through women. Two of Victoria's daughters were also carriers of the gene, and they passed it on to the royal families of Russia and Spain.

KEEP IT IN THE FAMILY

Heredity allows genes to survive for generations. The genes are passed down through families, and the characteristics they produce can be traced through the branches of a family tree (*right*). When people marry into a family and have children, they introduce new genes to the existing mix of genetic material. Scientists use family trees to determine the likelihood of serious diseases, such as breast cancer, being passed down to future generations.

Andrew Stevens (1897–1966) — Susan Kaplan (1905–1982)

Peter Smith (1860–1945) — Irene Roberts (1861–1953)

Lewis Stevens (1922–1980)

Robert Stevens (1923–1975)

Petra Stevens m. Maurice Smith (1928–) (1920–1984)

Rebecca Smith (1916–1990)

Darren Alexander m. Sarah Smith (1950–) (1954–)

Alex Smith (1960–)

Suzy Alexander (1979–)

Jack Alexander (1984–)

EXCEPTIONS TO THE RULE

The genders of most animals depend on the chromosomes they inherit from their parents. Some reptiles, however, such as crocodiles (*left*), are different. The sex of a crocodile is actually determined by the temperature of the egg it comes from. In the case of Australian saltwater crocodiles, an egg temperature of 88° to 89° Fahrenheit (31° to 32° Celsius) produces males. If the eggs are a few degrees cooler or warmer, they will hatch as females.

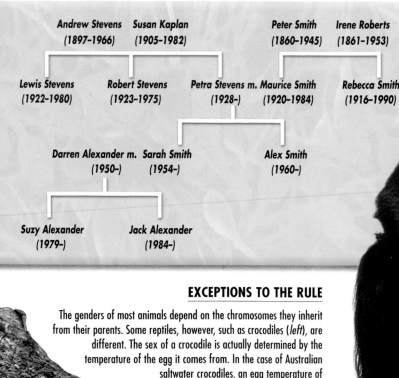

GINGER GENE

The "ginger gene," which is found in some people of European descent, is very old. Scientists at Oxford University in England compared the "ginger gene" with a similar gene found in chimpanzees. Through this comparison, they calculated that the "ginger gene" in humans could be up to 100,000 years old. Scientists believe modern humans migrated from Africa to Europe only about 40,000 years ago. The gene may have been passed down from *Neanderthals* (*left*), an older human species that already existed in Europe.

GREGOR MENDEL

The new science of genetics began in the 19th century with the work of an Austrian monk, Gregor Johann Mendel. He grew thousands of pea plants, collecting the seeds and using them to grow new plants. Mendel studied how often certain characteristics appeared in each generation. While Mendel knew nothing about chromosomes, genes, or DNA, he worked out the basic laws of heredity. One such law is that offspring inherit their characteristics equally from both parents.

GENETIC RELATIVES

Different species of plants and animals that are closely related to each other have a similar genetic code. For example, humans share up to 99 percent of their DNA with chimpanzees.

DOMINANT AND RECESSIVE GENES

Parents each pass on a copy of every gene they have to their children. The genes may be the same or different. When they are different, one gene may be dominant. A dominant gene hides the effect of a recessive gene. In humans, for example, the gene for brown eyes is dominant, while the gene for blue eyes is recessive. For a child to have blue eyes, both parents have to pass on the blue-eye gene, but for a child to have brown eyes, only one parent has to pass on the brown-eye gene.

CASE STUDY: SOLVING MYSTERIES

CRIME SCENE

Today, a crime scene will often involve forensic scientists, who gather evidence of a crime. They usually wear special overalls and gloves. These help ensure that the scientists do not drop their own clothes fibers, strands of hair, or flakes of skin onto the crime scene. The scientists do not want potential evidence to be contaminated with their own DNA, which could spoil the results of DNA tests. Outdoor crime scenes are sometimes covered with a tent to protect the area against wind and rain. The image shown above is taken from the British television drama *Silent Witness*, about the work of forensic scientists.

Today, many crimes and mysteries are solved using DNA testing. DNA left at the scene of a crime can be compared with DNA taken from suspects. If the DNA are the same, this match supports a case against the suspect. If the DNA are not the same, the test can prove that a suspect is innocent. DNA testing can also be used to clear up other controversies. A long-running mystery, involving a woman who claimed to be a member of the Russian royal family, was finally solved by a simple genetic test.

AN ANSWER ON ANASTASIA

During the Russian revolution of 1917, Tsar Nicholas II was removed from power, and he and his family were later killed. Rumors persisted, however, that at least one of the children, possibly a girl named Anastasia, had survived. In the 1920s, a woman calling herself Anna Anderson came forward claiming to be Anastasia. Research into her past, however, suggested she was actually a Polish woman named Franziska Schanzkowska. After Anna Anderson's death and cremation in 1984, genetic tests on tissue taken from her and from people related to the Russian royal family proved that she was not Anastasia. The 1956 movie *Anastasia* told the story of Anna Anderson, with the role of Anna played by Ingrid Bergman (*above*).

READING THE GENETIC CODE

Forensic scientists analyze biological samples to determine the samples' DNA. They use electrophoresis to analyze several samples of DNA at the same time. If the patterns of bands produced by the samples match exactly, then they probably came from the same person. If only some of the bands match, then the samples came from people who are probably related. If none of the bands match, then the samples are from people who are not related.

A GUILTY MAN?

Many samples of human tissue taken from past crime scenes are still kept today. In the case of crimes that are unsolved or where doubt exists about someone's guilt, these samples can be tested using new genetic techniques. James Hanratty was hanged in England in 1962 for the murder of scientist Michael Gregsten and the rape of Gregsten's companion, Valerie Storie. Nearly 30 years later, after a long campaign to clear Hanratty's name by those who believed he had been innocent, DNA tests were ordered. The results proved to be a shock for Hanratty's family and other supporters. They showed a match between DNA taken from Hanratty's body and traces on clothing found at the crime scene.

This is my son
JAMES HANRATTY
murdered by the state
for the A.6. murder

ELEVEN WITNESSES SWEAR HE
WAS IN RHYL 200 MILES AWAY
<u>WHEN</u> THE CRIME WAS COMMITTED

I DEMAND A
PUBLIC INQUIRY

AND JUSTICE
TO BE DONE

GENETIC PROFILING

In 1984, British geneticist Alec Jeffreys (*above*) developed a new technique called DNA profiling, which is also known as genetic fingerprinting. He noticed that some sequences, or series, of DNA are repeated over and over again within a person's genetic code. He also found that, with the exception of identical twins, everyone has a unique pattern of these repeated sequences. He then developed a way of making these patterns visible, so that scientists could compare the patterns from several samples of DNA.

FISHING FOR GENES

The genetic scientists pictured above are looking at a magnified image of chromosomes that have been treated so that parts of them glow. This technique is called FISH, which stands for Fluorescence In Situ Hybridization. It can be used to determine the base sequence of unknown DNA. FISH works by using short strands of known DNA, called probes. When the probes are mixed with the unknown DNA, they plug into the parts of the DNA that they fit. Scientists can see where the probes are because they have been treated with a chemical that makes them glow. Since the base sequence of the probes is known, scientists can determine the base sequence of the unknown DNA. By using different probes that plug into different parts of the unknown DNA, scientists can determine the entire base sequence of the unknown DNA.

BETTER SEQUENCING

Geneticists are developing new, faster ways of sequencing DNA. Electrophoresis, the process often used to analyze genetic material, involves passing an electric current through DNA, which heats it. Heat damages DNA, however, so early electrophoresis techniques had to work slowly to avoid a damaging buildup of heat. New techniques keep the DNA cool while using larger electric currents. The larger the current, the faster the process works.

CASE STUDY: THE HUMAN GENOME PROJECT

In 1988, scientists began working on one of the most important research projects in modern science. They set out to create a map of the human genetic code, which is also called the human genome. The Human Genome Project involves mapping, or sequencing, human DNA provided by several anonymous donors. The resulting DNA map will be typical of all human DNA. Forming a map of human DNA is not the same as understanding what DNA does. Scientists will need decades, and maybe even centuries, to learn what all of the different genes do and how they do it.

LISTING BASES

The first stage in the Human Genome Project has involved making a list of all the bases that make up the genes in human chromosomes. Since the human genetic code has about three billion bases, scientists perform this task with extremely fast DNA sequencers (*above*). Each short sequence of bases is worked out several times, so that any mistakes can be found and corrected. The next stage of the project will be much more difficult. It involves trying to understand exactly what the thousands upon thousands of human genes do.

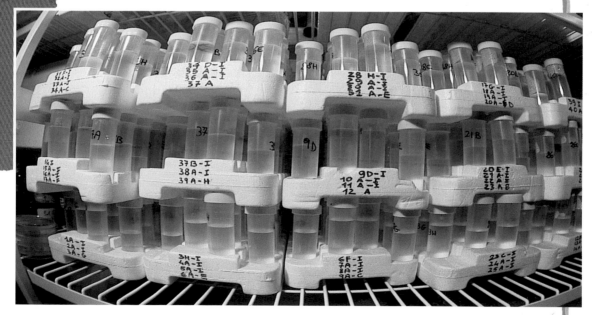

FINISHING THE MAP

The vials pictured above contain every gene in the human body. Scientists drew up the first rough map of the human genome in the beginning of 2001. This first rough listing, however, has some gaps in it that still have to be sequenced. The gaps are from parts of the DNA that probably contain few genes. Scientists hope to fill in all of these gaps so that they have the complete DNA sequence of each chromosome by 2003.

MAKING A CLONE

One technique for making clones is called nuclear transfer. First, scientists remove the nucleus of an unfertilized egg cell. Next, they fuse the egg cell with a cell from an adult animal by passing an electric current through the two cells. The cells become one cell, which then behaves like a fertilized egg and begins to divide. Finally, the cell is implanted into a female, where it develops normally into a baby.

adult sheep

Cells are removed from the adult sheep.

The new cell starts to divide like a normal cell.

One cell is fused with the egg.

The nucleus is removed from the egg.

A clone is born.

unfertilized egg

CLONING MAMMALS

The first attempts to clone animals, in the 1950s, involved inserting the nucleus of a frog cell into a frog egg. They were not successful. Frogs were cloned successfully for the first time in 1970, but the tadpoles did not develop into adult frogs. Scientists then turned their attention to cloning mammals, such as sheep and pigs. Five healthy piglets, Millie, Christa, Alexis, Carrel, and Dotcom (*right*), which were cloned from adult cells, were born on March 5, 2000.

CASE STUDY: CLONING

Cells or organisms that are genetically identical to each other are called clones. Clones can occur naturally — identical twins, for example, are natural human clones — but they can also be created in a laboratory. Creating clones can be risky, because many clones do not develop at all and other clones seem to have genes that don't work properly. Clones are very useful, however, in scientific research. For example, they enable scientists to test the effects of different chemicals and drugs on identical cells and organisms. It is possible to make human clones, but most people think cloning humans in laboratories is wrong.

DOLLY THE SHEEP

Dolly the sheep (*above*) made headlines in 1996 because she was the first mammal to be cloned from an adult cell. Animals had already been cloned using embryo cells, which can develop into all the different types of cells a new creature needs. Adult cells, however, become set in their ways. Dolly's creators found a way of making the adult cell behave like an embryo cell to produce all the different cells Dolly needed.

IDENTICAL BABIES

Identical humans occur when a fertilized egg splits, becoming more than one baby. The chances of identical babies being born decreases with the number of identical babies being born at the same time. Twins occur about every 250 births. Triplets occur about every 6,400 births. Four identical babies occur about every 500,000 births, while five identical babies only occur about every 41 million births!

ANIMAL BREEDING

People have been changing animal genes for thousands of years through a process called selective breeding, in which only animals with certain characteristics are chosen for breeding. Because selective breeding reinforces some genes and eliminates others, farmers often use it to improve certain qualities in their animals. Today, people like less fatty meat, so farmers have bred cattle and pigs that provide leaner cuts of meat. Genetics is also being used for the sake of conservation, helping to ensure that many endangered species continue to survive.

A MEATY ISSUE

Until about 300 years ago, cattle breeds developed naturally, but beginning in the 18th century selective breeding produced a variety of new breeds. The first of these was probably the meaty Hereford breed (*above*), which was produced by crossing white-faced Dutch cattle with small black Celtic cattle. At first, farmers bred the cattle naturally by putting selected bulls and cows together. Today, however, a technique called artificial insemination allows farmers to breed cattle without putting the bulls and cows together. This technique greatly increases the number of cattle that can be bred.

BREEDING WINNERS

The breeding of animals that are used for racing, such as dogs and horses, is strictly controlled and recorded, generation after generation. Champions are in great demand for breeding because their offspring have the best chance of also being champions. Thoroughbred horses (*above*) often have DNA profiles that are stored in an official register. A new DNA profile can be made at any time and compared to the original profile, to prove a horse's identity or to prove which horses are its parents.

LIFE SAVER

Endangered animals can sometimes be saved from extinction through breeding projects that try to increase their numbers. A good example of such a project involves the giant panda (*left*). Only about 1,000 giant pandas live in the wild, and about 120 live in zoos. The first giant panda that was bred using artificial insemination was born in 1963 at the Chengdu Giant Panda Breeding and Research Center in China. The center has produced 34 giant pandas using this method.

DOGS

Most modern dog breeds were produced artificially by selective breeding. The first dogs evolved from prehistoric wolves and jackals. About 6,500 years ago, only five types of dogs probably existed in the world — mastiffs, wolf-type dogs, sight hounds, pointing dogs, and herding dogs. Selective breeding between these types has produced all the modern breeds.

HELPING THE HAWKS

It is illegal in some countries to take hawks or their eggs from the wild. If the authorities are suspicious about a hawk owner's claim that the bird was bred in captivity, DNA profiling can prove or disprove the claim. If the bird was bred in captivity, its DNA will closely match the DNA of the birds from which it was bred, but if it was taken from the wild, its DNA will not match. In 1993, a hawk breeder in Scotland was convicted of "nest theft," which was proven by DNA profiling.

SUPERPLANTS

Genetic engineering can produce new varieties of foods, such as tomatoes that stay ripe for longer periods of time without rotting. It can also produce new varieties of crops that are more resistant to disease. This resistance could prevent disasters such as the Irish potato famine of the 1840s. During the famine, more than a million people starved and even more left the country for a new life in North America. The famine was caused by the failure of the country's main crop, potatoes, which were ruined by a fungus disease called potato blight.

REMARKABLE RICE

About four billion people in the world depend on rice (*above*) as a food source, so it is a major crop. Rice was the first major food crop to have its genome mapped, and scientists are modifying that genome to develop new varieties of rice. Many people in poor countries have low levels of vitamin A and iron, which can lead to blindness and other health problems. A new variety of rice has more vitamin A and iron.

PUTTING NEW GENES INTO PLANTS

Scientists have several ways of putting new genes into plants. Bacteria known to infect the plants can be used to carry the DNA. The genes can also be stuck to metal particles that are then fired into the plant cells like bullets. Other methods include using electric shocks to break down plant cell walls to let in new genes, or injecting new genes directly into the plant cells through a needle.

USING BACTERIA

DNA to be inserted

USING GENE BULLETS

Bacteria carries the DNA in a ring called a plasmid.

Metal particles are coated with DNA.

plant cells

The bacteria attacks the plant cell, "infecting" it with the DNA.

The particles are fired into the plant cells.

In soil, a plantlet grows into a genetically adapted plant.

The altered plant cell begins to divide.

Cells are regenerated into plantlets.

GENETICS & PLANTS

Modern crops were developed from smaller wild plants by selective breeding. Farmers made sure that the best plants pollinated each other, generation after generation. Through selective breeding, farmers improved their crops by reinforcing certain characteristics, such as the size and

shape of a plant's fruit. Scientists of today, however, have introduced a new way of improving crops by changing plant DNA in the laboratory. This process is called genetic engineering. Through genetic engineering, scientists can change plants, such as barley and corn, so that they can better resist diseases, pests, and lack of water. Scientists are also creating plants that can grow in a wider range of soil types and can produce fruit that lasts longer without rotting. Crops that have been genetically altered are known as genetically modified, or GM, crops.

WHEAT TO EAT

Grains such as wheat are major crops. Since a lot of the world's agriculture relies on only five types of grain, geneticists are trying to map and understand the genomes of these types so they can be modified to resist pests, disease, or a changing climate. Geneticists hope to develop more resilient strains of wheat in the future.

THE GREAT GENE HUNT

Scientists are constantly checking plants from all over the world for genes that might be used to modify other plants. Any plant might have a gene that could improve the qualities of another plant. Scientists are also concerned about saving rare plants from extinction, because any useful genes they contain would then be lost forever.

GENE THERAPY

Today, scientists can read the genetic code inside human cells and can actually change that code through genetic engineering. About 3,000 medical conditions in humans are caused by inheriting faulty genes. The treatment of these conditions through the use of genetic engineering is called gene therapy. The 19th century was dominated by engineers who worked with iron and steel. The 20th century was dominated by electronic engineers who developed the computers and other electronic equipment that we now use. The 21st century may be dominated by the work of genetic engineers and gene therapists.

GERMLINE THERAPY

Most gene therapy affects only the patient being treated. The treatment does not affect the sex cells, so it is not passed on to any children the patient might later have. In 1999, a team of Canadian scientists succeeded in giving a mouse an extra chromosome that was passed on to the mouse's offspring. If this technique was to be used in humans, any genetic change would be passed on to future generations. The technique is called germline therapy.

ALTERING VIRUSES

A virus that will be used to carry a gene into a human cell is first made safe by removing some of its genetic material. A healthy human gene is then introduced into the virus, becoming part of the virus's own genetic code. Trillions of modified viruses are injected into the patient. Each virus attaches itself to a human cell, sending its genetic material into the cell to become part of the cell's DNA.

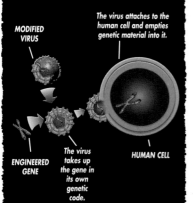

MODIFIED VIRUS

The virus attaches to the human cell and empties genetic material into it.

ENGINEERED GENE

The virus takes up the gene in its own genetic code.

HUMAN CELL

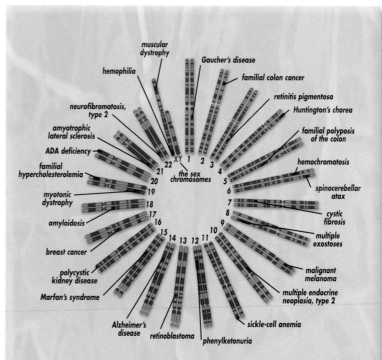

GENETIC DISEASES

Geneticists have found the locations of many of the genes that cause particular inherited medical conditions. This diagram shows a complete set of the 23 pairs of human chromosomes and in which genes the conditions occur. Geneticists located these genes by studying families that had several relatives suffering from the same condition. The chromosomes of these families were compared to those taken from healthy families. The differences showed where the faulty genes were located.

GETTING IN

New DNA can be injected directly into a cell using a thin needle, but better methods exist. The cells can be treated with chemicals that make their membranes "leaky," so that the new genes can pass through into the nucleus. The most common methods for getting DNA into cells, however, involve using viruses and bacteria.

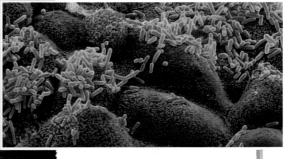

BACTERIA BENEFITS

Bacteria can be used as microscopic factories to create all kinds of chemicals that can help a patient. To get bacteria to create a certain chemical, scientists add a gene that produces the chemical to the bacteria's own genes. This technique allows hormones, insulin (for treating diabetes), and a variety of proteins to be produced. Without genetically engineered bacteria, these substances would have to be extracted from other people, which involves a risk of infectious diseases spreading from the donor to the patient.

CYSTIC FIBROSIS

Cystic fibrosis is a common inherited illness. It mainly affects the lungs and makes breathing difficult. It is caused by a single faulty gene on chromosome 7. The gene is recessive, so two faulty copies must be inherited — one from each parent — to cause the illness. It is most common in Europe, where it affects 1 in 3,000 people. Because cystic fibrosis involves only one gene, it is potentially one of the simplest diseases to treat genetically. Once a system has been developed to deliver a healthy gene, the disease should be conquered. Scientists are experimenting with treatments that involve using viruses to carry new, healthy genes directly into the lungs.

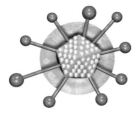

VIRUS CARRIERS

Viruses such as the adenovirus (*left*) are very good at getting inside cells and combining their genetic material with the material that already exists. They are ideal "carriers" to take engineered genes into cells. Scientists often use the adenovirus, which normally causes a coldlike infection, as a carrier.

DESIGNER BABIES?

In the future, parents might be able to select certain characteristics they want their children to have. Some people fear that if this becomes possible, wealthy parents will be able to pay for "designer babies," ensuring that their children have the genetic makeup for the best chances in life. Since less wealthy parents might not be able to afford designer babies, the result could be a "genetic lower class" of people with more faulty genes than wealthier people. Most geneticists are extremely opposed to using genetic engineering to create designer babies.

gene tiX
snowball
0161 834 020

SICKLE CELLS VS. MALARIA?

Human blood cells are normally disk shaped, but some people have blood cells that are shaped like a sickle, or crescent. People who inherit two copies of the sickle-cell gene suffer from a serious condition called sickle-cell anemia. People who inherit one copy do not have enough sickle cells to cause a problem, and the sickle cells protect against malaria. If gene therapy were to reduce the number of people with one copy of the sickle-cell gene in an area where malaria is common, more people in that area could be susceptible to malaria.

THE RISKS OF GENETIC ENGINEERING

Genetic engineering has tremendous potential for improving our lives, but many people are worried about its use. For example, they are concerned that altered genes released into the environment may behave in unpredictable ways, and that this new technology could lead to disaster. Most geneticists, however, are responsible scientists who recognize the risks of genetic engineering and are trying to address them.

GENETIC TESTS?

Genetic tests can now detect a variety of inherited illnesses that a person could possibly develop, including muscular dystrophy, cystic fibrosis, and hemophilia. In most cases, these tests are very useful because they can lead to early detection and treatment. The tests can, however, raise some troubling issues. For example, parents could find out that their child may develop a disease for which no effective cure exists.

GM CROPS: GOOD OR BAD?

Some people who are opposed to the genetic engineering of plants have destroyed genetically modified crops. They insist that regular crops can be made more productive by improving the soil, eliminating pests, and persuading governments to spend less on weapons and more on agriculture. Scientists, however, believe GM crops are necessary to help feed people in poor countries.

GENETIC DISASTER

In 1999, geneticists were reminded that gene therapy carries risks like any other treatment when a patient receiving gene therapy died. Jesse Gelsinger, an 18-year-old man from Arizona, suffered from a rare inherited liver disorder called OTC deficiency. He was being treated with adenoviruses altered to carry healthy OTC genes into his liver. When trillions of the viruses were injected into his body, he suffered a violent reaction to them and died. Unfortunately, no person has yet been cured of a disease using gene therapy.

GENETICS TODAY & TOMORROW

Genetic research and engineering has just begun. No one knows for sure how genetics will affect our species in the next hundred years, but our new knowledge of how the human body works seems certain to completely change our understanding of diseases and how to treat them. Gene therapy holds the promise of cures for many diseases, including cancer, while the genetic engineering of plants could put an end to famines that result in the deaths of millions of people. We may be on the brink of a scientific and medical revolution that will be truly amazing.

THE BODY SHOP

Stem cells are "general-purpose" cells that multiply easily and develop into specialized cells, such as those that make up skin, blood, muscles, and internal organs. Scientists hope they can eventually use stem cells to produce human organs. Modified stem cells pass on their new genes to every new cell, but scientists do not yet know how to make them change into the particular types of cells that are needed.

GROWING CORNEAS

The cornea is a transparent "window" at the front of the eye. It is a complicated structure with three layers of cells. In 2000, scientists succeeded in growing an artificial cornea that worked like a real one. This artificial cornea will be used in testing the safety of medicines, reducing the number of tests done on animals.

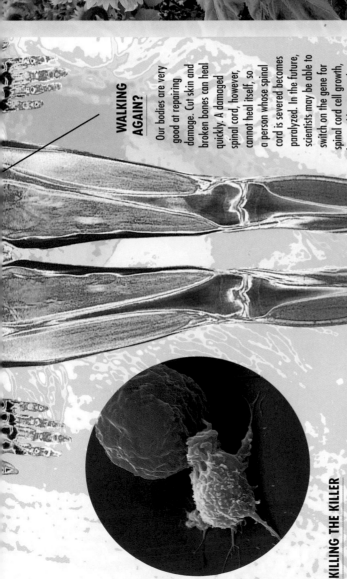

AN OIL FIELD?

The fuel we burn in our cars is made from oil. This oil is pumped out of underground oil fields, but one day these fields will run dry. In the future, we may be driving cars powered by plant fuel. Plants such as sunflowers (*above*) may be genetically modified to produce an oil that will be just as effective as the fuel we use today.

WALKING AGAIN?

Our bodies are very good at repairing damage. Cut skin and broken bones can heal quickly. A damaged spinal cord, however, cannot heal itself, so a person whose spinal cord is severed becomes paralyzed. In the future, scientists may be able to switch on the gene for spinal cord cell growth, thus enabling injured spinal cord tissue to repair itself.

KILLING THE KILLER

Cancer is caused by cells multiplying out of control. They form a growth called a tumor (*above, left*). Tumor cells that are carried throughout the body by the blood system can start new tumors growing in other places. Because the cells belong to the body and are not alien invaders, the body's immune system does not attack them. Some families carry a single faulty gene that places family members at increased risk for certain types of cancer. In the future, gene therapy may be able to correct the fault and reduce the risk of cancer in these families.

GLOSSARY

amino acids: simple chemical compounds that join together inside cells to make proteins.

cell: the smallest unit of life that can function on its own.

chromosome: a short length of DNA inside a cell's nucleus.

clones: organisms that are genetically identical.

cytoplasm: a jellylike material surrounding the nucleus inside a cell.

DNA: the abbreviation for deoxyribonucleic acid, which is the chemical in cells that contains genetic instructions.

DNA sequencer: a device that allows geneticists to determine the sequence, or order, of chemical units in DNA.

dominant gene: a gene that can produce a characteristic in an organism with a copy from just one parent.

double helix: the shape of the DNA molecule, which resembles a twisted ladder.

electrophoresis: a process that allows geneticists to create special patterns from DNA samples, so that the samples can be compared.

endoplasmic reticulum: in a cell, a network of folded membranes where proteins are made.

FISH: the acronym, or initials, for Fluorescence In Situ Hybridization, a process that allows geneticists to determine the sequences of unknown DNA.

genes: the parts of a chromosome that determine the characteristics of an organism.

Golgi complex: a structure inside a cell that stores and distributes chemicals.

human genome: the genetic code for humans, which consists of all the genes in the 23 pairs of chromosomes found in human cells.

mitochondria: structures inside cells that release energy to power chemical reactions.

molecule: a group of atoms, which are the tiny building blocks that make up all substances.

mutation: a change in a cell's DNA.

natural selection: the process in which a species, over a very long period of time, changes so it has characteristics that are best suited for its survival.

nucleus: the control center of a living cell, where most of its DNA is contained.

proteins: molecules found in all living cells that are involved in many of the jobs that cells perform, and which are each made up of a long chain of chemical compounds called amino acids.

recessive gene: a gene that can produce a characteristic in an organism only when copies from both parents are present.

ribosome: a ball-shaped structure inside a cell that makes proteins according to instructions from the DNA in the nucleus.

SRY gene: the gene in a human Y chromosome that is responsible for making a baby a male.

virus: a tiny particle of genetic material that can cause disease in plants and animals and can only multiply when it is inside a living cell.

MORE BOOKS TO READ

Cloning: Frontiers of Genetic Engineering. Megatech (series). David Jefferis, Davies Jefferis (Crabtree Publishers)

Gene Therapy. Great Medical Discoveries (series). Lisa Yount (Lucent Books)

Genetic Engineering: Debating the Benefits and Concerns. Issues in Focus (series). Karen Judson (Enslow Publishers)

Genetics. Science Fact Files (series). Richard Beatty (Raintree/Steck-Vaughn)

Gregor Mendel: And the Roots of Genetics. Oxford Portraits in Science (series). Edward Edelson (Oxford University Press)

How the Y Makes the Guy. Microexplorers (series). Patrick A. Baeuerle, Norbert Landa (Barrons)

WEB SITES

Cracking the Code of Life.
    www.pbs.org/wgbh/nova/genome/

DNA from the Beginning.
    www.dnaftb.org/dnaftb/

Genetics Science Learning Center.
    http://gslc.genetics.utah.edu/

I Can Do That!
    www.eurekascience.com/ICanDoThat/

INDEX

ACKNOWLEDGEMENTS

The original publisher would like to thank Advocate, Oliver Zaccheo from the
Department of Genetics at the University of Leicester, and Elizabeth Wiggans for their assistance.

Picture Credits: t=top, b=bottom, c=center, l=left, r=right
BBC Photo Library: 18tl. Corbis Images: 4c, 5b, 10tl, 19r, 23br, 31tr. Corbis Stockmarket: 30t. Environmental Images: 33t Kobal Collection: 18/19c.
Mary Evans Picture Library: 16tl. Pictor: 12/13t. Popperfoto: 22/23c. Science Photo Library: 4tl, 5t, 6cl, 6b, 6/7c, 7t, 8b, 9t, 9b, 11t, 11r, 12b, 13t, 13c,
14b, 15t, 16t, 16b, 17b, 18b, 20t, 21t, 21b, 23cr, 24t, 24c, 24br, 25cl, 25r, 26l, 26c, 27r, 27b, 28t, 29t, 29r, 32/33c, 33b. Tony Stone Images: 8l, 8/9c,
16/17b. Still Pictures: 30/31.

Every effort has been made to trace the copyright holders, and we apologize in advance for any unintentional errors or omissions.